Online Searching on STN

BEILSTEIN
Workshop Manual

Springer

© 1989 by Springer-Verlag Berlin Heidelberg
Originally published by Springer-Verlag Berlin Heidelberg New York in 1989
Softcover reprint of the hardcover 1st edition 1989
Quoting or copying of material from this publication for educational purposes is encouraged, providing acknowledgment is made of the source of such material.

9 8 7 6 5 4 3 2 1

ISBN 978-1-4684-7988-1   ISBN 978-1-4613-9635-2 (eBook)
DOI 10.1007/978-1-4613-9635-2

Contents

I. Introduction

    What Is the BEILSTEIN Database? I-2
    A Handbook Entry and the Corresponding Online Record I-3
    Features of the BEILSTEIN Database on STN I-6

II. Substance Searching

    Substance Names II-2
    Searching Substance Names in the Basic Index II-5
        Proximity II-6
        Allowing for Spelling Variations II-9
        Search Example II-10
    Special Fields for Substance Name Searching II-11
    Molecular Formula Searching II-15
    Element and Atom Counts II-16
        Search Example II-17
    Structure Searching II-18
        Search Examples II-19
    Tautomers II-21
    General Conventions for Representing Substances II-22
    Using Other Substance Identifiers II-29
    Similarity Searching II-34
        Search Example II-36
    Displaying Substance Information II-37

III. Searching for Physical Property Information

    Types of Property Information Available III-2
    General Information of Property Searching III-3
    Searching for a Property—Search Examples III-6
    Searching for a Specific Data Value or Values III-8
        Search Examples III-12
    Displaying Property Data III-14

IV. Searching for Preparations and Reactions

    General Information IV-2
    Searching for Preparations IV-3
        Search Examples IV-4
    Searching for Reactions IV-7

Searching for Reactions of the Type A + B IV-8
    Search Examples IV-9
Searching for Reactions of the Type A → B—
    Search Examples IV-10
Displaying Reaction Information IV-11

V. Other Search Information

    Additional Search Hints V-2
    Searching for References V-4
    System Limits V-5
    Help V-6

VI. Practice Questions VI-2

VII. Appendices

    A. Periodic Group Codes VII-2
    B. Atomic Weights VII-3
    C. Controlled Terms VII-4

I. Introduction

INTRODUCTION I-2
What Is the BEILSTEIN Database?

- The BEILSTEIN database contains records for
  - Organic substances cited in the Beilstein Handbook, Basic Series and Supplementary Series I-IV, covering the time period 1830-1960.
  - Organic substances from the Beilstein collection of excerpts from the primary literature that have not yet been critically reviewed, covering the time period 1960-1980.

- The substances in the database include stereoisomers, salts, and multicomponent systems.

- The substances in the database have known, verified constitutions, are pure, and have factual data associated with them.

- Substance identification information available for search and display in the database includes
  - Beilstein Registry Numbers
  - Substance names
  - Molecular formulas
  - Structure diagrams
  - Lawson Numbers

- Essentially all of the data from the Handbook are available in the database, e.g.,
  - Physical properties
  - Isolation, preparation, purification information
  - Characterization information
  - Chemical behavior and reactions

- Data for the unevaluated substances consists of specific numeric values for a few properties and keywords for other types of information.

- References to the original literature are included in the database.

- Information appearing in the records in the database is selected from journals, patents, and monographs.

- Most of the information in the database is in English.

INTRODUCTION   I-3
A Handbook Entry and the Corresponding Online Record

SO 4-17-00-04983

4. Oxo-Verbindungen $C_{11}H_{12}O_2$.

    1. *4-Oxo-2-phenyl-tetrahydropyran, 2-Phenyl-tetrahydropyron-(4)*

$C_{11}H_{12}O_2 = \begin{matrix} H_2C \cdot CO \cdot CH_2 \\ | \qquad\qquad\quad | \\ H_2C - O - CH \cdot C_6H_5 \end{matrix}$ *B.* Neben anderen Produkten bei der Hydrierung von 2-Phenyl-pyron-(4) bei Gegenwart von kolloidem Palladium in verd. Alkohol (BORSCHE, PETER, A. 453, 158). — Wurde als 4-Phenyl-semicarbazon isoliert.

    4-Phenyl-semicarbazon $C_{18}H_{19}O_2N_3 = C_6H_5 \cdot C_5H_7O(:N \cdot NH \cdot CO \cdot NH \cdot C_6H_5)$, Kristalle (aus verd. Alkohol), F: 195–196° (BORSCHE, PETER, A. 453, 158).

SO 2-17-00-00346

(±)-**2-Phenyl-tetrahydro-pyran-4-on** $C_{11}H_{12}O_2$, Formel IV (X = H) (E II 346).

    B. Beim Behandeln von opt.-inakt. 2-Phenyl-tetrahydro-pyran-4-ol mit Kaliumdichromat und wss. Schwefelsäure (*Cornubert et al.*, Bl. 1954 534; *Hanschke*, B. 88 [1955] 1053, 1059).

    Kp_{23}: 170—175° (*Co. et al.*); $Kp_{0,5}$: 120—122°; $Kp_{0,1}$: 102—104° (*Ha.*). D_4^{20}: 1,1138; n_D^{20}: 1,5410 (*Ha.*).

(±)-**2-Phenyl-tetrahydro-pyran-4-on-oxim** $C_{11}H_{13}NO_2$, Formel V (X = OH).

    B. Aus (±)-2-Phenyl-tetrahydro-pyran-4-on und Hydroxylamin (*Cornubert et al.*, Bl. 1954 534).

    Krystalle (aus A.); F: 125—126°.

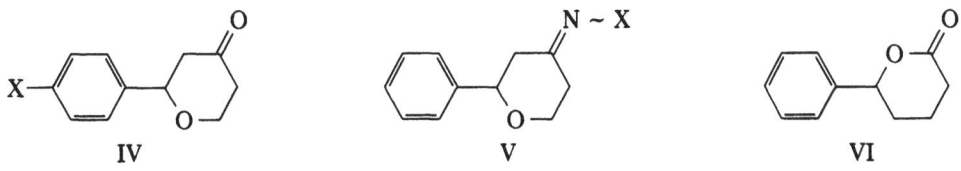

    IV   V   VI

INTRODUCTION I-4
A Handbook Entry and the Corresponding Online Record

```
BRN   141748   Beilstein
MF    C11 H12 O2
CN    2-phenyl-tetrahydro-pyran-4-one
      2-Phenyl-tetrahydro-pyran-4-on
FW    176.21
SO    4-17-00-04983; 2-17-00-00346
LN    17942
```

[structure of 2-phenyl-tetrahydro-pyran-4-one]

---> Handbook Data <---
Isolation from Natural Product:
INP   wurde als 4-Phenyl-semicarbazon isoliert
      Reference(s):
      1. Borsche, Peter, Liebigs Ann.Chem. 453 ⟨1927⟩, 158, CODEN: LACHDL

---> Handbook Data <---
Preparation:
PRE
      Educt: 2-phenyl-pyrone-(4)
      Reag: palladium, diluted alcohol
      Detail: Hydration
      Reference(s):
      1. Borsche, Peter, Liebigs Ann.Chem. 453 ⟨1927⟩, 158, CODEN: LACHDL
PRE
      Educt: 2-phenyl-tetrahydro-pyran-4-ol
      Reag: potassium dichromate, aqueous sulfuric acid
      Reference(s):
      1. Hanschke, Chem.Ber. 88 ⟨1955⟩ 1053, 1059, CODEN: CHBEAM
      2. Cornubert et al., Bull.Soc.Chim.Fr. 1954 534, CODEN: BSCFAS

INTRODUCTION I-5
A Handbook Entry and the Corresponding Online Record

```
---> Handbook Data <---
Boiling Point:
Value (BP)                 Press. (BP.P)             Ref.    Note
(Cel)                      (Torr)
------------------------+-------------------------+------+------
170.00-175.00              23                        1
120.00-122.00              0.5                       2
102.00-104.00              0.1                       2

Reference(s):
1. Cornubert et al., Bull.Soc.Chim.Fr. 1954 534, CODEN: BSCFAS
2. Hanschke, Chem.Ber. 88 ⟨1955⟩ 1053, 1059, CODEN: CHBEAM

---> Handbook Data <---
Density (liquid):
Value (DEN)     Temp. (DEN.T)    Ref. Temp.       Ref.   Note
(g/cm**3)       (Cel)            (Cel)
--------------+----------------+----------------+------+------
1.11380         20.0             4.0              1

Reference(s):
1. Hanschke, Chem.Ber. 88 ⟨1955⟩ 1053, 1059, CODEN: CHBEAM

---> Handbook Data <---
Refractive Index:
Value (RI)      Wavel. (RI.W)    Temp. (RI.T)     Ref.   Note
(—)             (nm)             (Cel)
--------------+----------------+----------------+------+------
1.54100         589.00           20.0             1

Reference(s):
1. Hanschke, Chem.Ber. 88 ⟨1955⟩ 1053, 1059, CODEN: CHBEAM
```

INTRODUCTION I-6
Features of the BEILSTEIN Database on STN®

- Search features

    —Substructure, numeric, and text searching is available.
    —Complete recall of numeric data, including data specified as numeric ranges, e.g.,

    => S 120 / MP

    retrieves

    MP 120 deg. C
    MP 119.1 – 123.3 deg. C
    MP > 100 deg. C

    —Easy coordination of properties and the parameters associated the measurement, e.g.,

    BP—Boiling point
    BP.P—Pressure at which the boiling point was measured

- Display features

    —Each record contains the information from all volumes of the handbook.
    —Structure diagrams are available for display online and in offline prints.
    —Numeric data are displayed in tabular form for easy scanning.
    —Default display format includes the substance identification information, *plus* the display fields that contain the hit terms.
    —HIT display format for displaying only the fields containing the search terms.
    —DISPLAY FA for listing all of the display fields in a specified answer.
    —Predefined display formats for displaying specific subsets of data, e.g., PHYS for displaying physical data.

- HELP messages supply information on search and display fields and units for numeric data.

II. Substance Searching

Substance Name Searching
Molecular Formula Searching
Structure Searching
General Conventions
Using Other Substance Identifiers

SUBSTANCE SEARCHING   *II-2*
Substance Names

- Chemical substance names are available for all substances from the Handbook and are displayed in the CN field.

  All substance names in the CN display field are in English.

  ```
  BRN   235077   Beilstein
  CN    2-Benzoyl-2,3-epoxy-3-phenyl-propionic acid-amide
        2-Benzoyl-2,3-epoxy-3-phenyl-propionsaeure-amid
  ```

- Additional synonyms are available for many substances, and these are displayed in the SY field.

  Substance names in the SY display field may be in German or English.

  ```
  BRN   385526   Beilstein
  CN    N,N'-⟨4-phenylazo-m-phenylene⟩-bis-phthalimide
        N,N'-⟨4-Phenylazo-m-phenylen⟩-bis-phthalimid
  SY    2,4-Diphthalimido-azobenzol
  ```

- Most substances from the unreviewed excerpts from the primary literature have substance names, and these names are displayed in the SY field.

SUBSTANCE SEARCHING
Substance Names

- Substance names are available for chemical derivatives of Beilstein Registry substances

  Searchable substance names in the CDER display field are in English.

```
BRN   384151   Beilstein

--- > Handbook Data < ---
Chemical Derivative :
CDER 2,4-dinitro-phenylhydrazone (Mp.: 143 degree-144 degree)
      Reference(s):  1. Speroni, Giachetti, Gazz.Chim.Ital. 83 ⟨1953⟩ 192, 209, CODEN:
         GCITA9
```

SUBSTANCE SEARCHING *II-4*
Substance Names

- Substance names are available for educts and reagents used to prepare Beilstein Registry substances, and for by-products of these reactions.

  Searchable substance names in the PRE display field are in English.

```
BRN   385591   Beilstein

---> Handbook Data <---
Preparation:
PRE
    Educt:    2,5-dimethyl-4-pyrrolidinomethyl-phenol
    Reag:     benzene, sodium, isopropyl alcohol
    Detail:   Anschliessend Behandeln mit Thiophosphorylchlorid.
    Reference(s):  1. Pat. No.: 2 759 961, US, Fitch, 1952
```

- Substance names are available for reaction partners, reagents, and products of reactions of Beilstein Registry substances.

  Searchable substance names in the REA display field are in English.

```
BRN   385536   Beilstein

---> Handbook Data <---
Chemical Reaction:
REA
    Part.:    dioxane, methanol. caustic potash solution
    Detail:   Ansaeuern einer wss. Loesung des Reaktionsprodukts
    Prod.:    5,5'-bis-carboxymethylene-5H,5'H-⟨2,2'⟩bithienylidene
    Reference(s):  1. Steinkopf, Hanske, Liebigs Ann.Chem. 541 ⟨1939⟩ 238, 252,
                      CODEN: LACHDL
```

SUBSTANCE SEARCHING
Searching Substance Names in the Basic Index

- Name segments from all chemical substance names are posted in the Basic Index.
- Names segments are created by parsing names at spaces and punctuation.
- Additional smaller chemically significant substance name segments may be available for names segments from the CN display field; e.g., DIMETHYL, DI, and METHYL are posted for the segment DIMETHYL when it appears in the CN display field.

Substance Name	Name Segments in /BI
7-Hydroxy-4,6-dimethyl-phthalide	7
	4
	6
	hydroxy
	dimethyl
	phthalide

SUBSTANCE SEARCHING
Searching Substance Names in the Basic Index

- (W) —Name segments adjacent in the order specified.
  (nW)—Name segments n or fewer terms apart and in the order specified.

```
=> S DICHLORO (W) PHENYL
        10830 DICHLORO
       100166 PHENYL
L1         724 DICHLORO (W) PHENYL

=> D 1-2 HIT

L1  ANSWER 1 OF 724

---> Handbook Data <---
Preparation:
PRE
    Educt:  N-⟨2,5-dichloro-phenyl⟩-anthranilic acid
    Reag:   sulfuric acid
    Reference(s):  1. Nisbet, J.Chem.Soc. 1933 1372, CODEN: JCSOA9
```

- (A) —Name segments adjacent in any order.
  (nA)—Name segments n or fewer terms apart and in any order.

```
=> S DICHLORO (A) PHENYL
        10830 DICHLORO
       100166 PHENYL
L3         739 DICHLORO (A) PHENYL

=> D 274 HIT
L3  ANSWER 274 OF 739

---> Handbook Data <---
Preparation:
PRE
    Educt:  phenyl-⟨4-nitro-phenyl⟩-dichloro-methane, silver azide
    Reag:   diisopentyl ether
    Reference(s):  1. Schroeter, Chem.Ber. 42 ⟨1909⟩, 3360, CODEN: CHBEAM
```

Searching Substance Names in the Basic Index

- (S)—Name segments anywhere in the same name.

```
=> S DICHLORO (S) PHENYL
         10830 DICHLORO
        100166 PHENYL
L8        1693 DICHLORO (S) PHENYL

=> D 1618 HIT

L8   ANSWER 1618 OF 1693
CN   ***3,5-dichloro-4,6-dioxo-1-phenyl-1,4,5,6-tetrahydro-pyridine-2-carboxylic
     acid***
     3,5-Dichlor-4,6-dioxo-1-phenyl-1,4,5,6-tetrahydro-pyridin-2-carbonsaeure

---> Handbook Data <---
Chemical Reaction:
REA
     Part.:  by heating upon the melting point
     Prod.:  carbon dioxide,
             1-phenyl-3.5-dichloro-2.4-dioxo-1.2.3.4-tetrahydro-pyridine
     Reference(s):  1. Zincke, Fuchs, Liebigs Ann.Chem. 267 ⟨1892⟩, 29, CODEN:
         LACHDL
REA
     Part.:  acetic acid anhydride, sodium acetate
     Prod.:  1-phenyl-3.5-dichloro-4(or 2)-acetoxy-pyridon-(2 or 4)
     Reference(s):  1. Zincke, Fuchs, Liebigs Ann.Chem. 267 ⟨1892⟩, 29, CODEN:
         LACHDL
```

Searching Substance Names in the Basic Index

- (L)—Name segments anywhere in the same type of field.

```
= > S DICHLORO (L) PHENYL
            10830 DICHLORO
            100166 PHENYL
L9          2064 DICHLORO (L) PHENYL

= > D 881 HIT

L9   ANSWER 881 OF 2064

---> Handbook Data <---
Preparation:
PRE
    Educt:   3,6-dichloro-pyridazine, sulfanilamide
    Reag:    K2CO3, NaCl
    Reference(s):  1. Druey et al., Helv.Chim.Acta 37 〈1954〉 121,133, CODEN: HCACAV
    2. Pat. No.: 2671086, US, Am.Cyanamid Co., 1951
PRE
    Educt:   〈4-(6-chloro-pyridazin-3-ylsulfamoyl)-phenyl〉-carbamic acid ethyl ester
    Reag:    aq. NaOH solution
    Reference(s):  1. Druey et al., Helv.Chim.Acta 37 〈1951〉121,133, CODEN: HCACAV
    2. Pat. No.: 951929, DE, CIBA, 1953
```

Searching Substance Names in the Basic Index

Allowing for Spelling Variations

- Use ? or # to allow for variation at the end of a name or name segment:

  THIOSEMICARBAZON#   retrieves   THIOSEMICARBAZON
  THIOSEMICARBAZONE

  ISONICOTINOYL?   retrieves   ISONICOTINOYL
  ISONICOTINOYLHYDRAZONE
  etc.

- Use ! if unsure of an internal character:

  DISAMINYL!THER   retrieves   DISAMINYLATHER
  DISAMINYLETHER

  NOTE: The German ä, ü, and ö are represented as ae, ue, and oe in the online record.

SUBSTANCE SEARCHING
Searching for Substance Names in the Basic Index

Search Example

QUESTION: Find information on methyl derivatives of isocoumarin.

```
=> S (METHYL OR DIMETHYL OR TRIMETHYL OR TETRAMETHYL) (S)
(ISOCOUMAR? OR ISOCUMAR?)
            153943  METHYL
             45598  DIMETHYL
              9618  TRIMETHYL
              4068  TETRAMETHYL
               195  ISOCOUMAR?
                97  ISOCUMAR?
                42  (METHYL OR DIMETHYL OR TRIMETHYL OR TETRAMETHYL
L10             42  (METHYL OR DIMETHYL OR TRIMETHYL OR TETRAMETHYL

=> D 1,4,14 HIT

L10 ANSWER 1 OF 42

CN      ***4,6,8-triacetoxy-5-methoxy-3-methyl-isocoumarin***
        4,6,8-Triacetoxy-5-methoxy-3-methyl-isocumarin

L10 ANSWER 4 OF 42

CN      ***6,7-dimethyl-3,4-diphenyl-isocoumarin***
        6,7-Dimethyl-3,4-diphenyl-isocumarin

L10 ANSWER 14 OF 42

---> Handbook Data <---
Preparation:
PRE
    Educt:  5.7-diethoxy-4-methyl-isocoumarin, ammonia
    Reag:   alcohol
    Temp:   100.0 Cel
    Reference(s):  1. Pat. No.: 73700, D.R.P., Fritsch Friedlaender: 3,970
```

Special Fields for Substance Name Searching

- /CN   Chemical name for Beilstein Registry substances, including synonyms, when available.

    Names in /CN are posted as bound phrases, including punctuation.

    Reserved messenger characters must be masked in search terms in /CN. The reserved characters are

    $$/ \; (\;) \; < \; > \; ? \; ! \; \# \; " \; ' \; =$$

    Useful for: Searching for information about a specific file substance, e.g.,

```
=> S PYRIDINE / CN
L3              1 PYRIDINE / CN

=> D HIT

L3 ANSWER 1 OF 1

CN        ***pyridine***
          Pyridin
```

SUBSTANCE SEARCHING
Special Fields for Substance Name Searching

- /CNS   Chemical name segments from the names in the /CN search field.

    The chemical name segments are created by parsing the name at spaces and punctuation.

    Additional smaller name segments may be available for name segments from the CN display field.

    Useful for: Name segment searching for file record substances, e.g.,

```
=> S (DIMETHOXY (S) DIMETHYL (S) PHTHALID#) / CNS
         10281 DIMETHOXY / CNS
         34052 DIMETHYL / CNS
          1779 PHTHALID# / CNS
L1           6 (DIMETHOXY (S) DIMETHYL (S) PHTHALID#) / CNS

=> D 1-3 HIT

L1   ANSWER 1 OF 6

CN      ***3-(6-⟨2-(dimethyl-oxy-amino)-ethyl⟩-4-methoxy-benzo⟨1,3⟩dioxol-
        5-ylmethyl)-6,7-dimethoxy-phthalide***
        3-(6-⟨2-(Dimethyl-oxy-amino)-aethyl⟩-4-methoxy-benzo⟨1,3⟩dioxol-5-yl
        methyl)-6,7-dimethoxy-phthalid

L1   ANSWER 2 OF 6

CN      ***3-⟨6,8-dimethyl-⟨2⟩quinolylmethyl⟩-6,7-dimethoxy-phthalide***
        3-⟨6,8-Dimethyl-⟨2⟩chinolylmethyl⟩-6,7-dimethoxy-phthalid

L1   ANSWER 3 OF 6

CN      ***3-⟨3,7-dimethyl-octa-2,6-dienyloxy⟩-6,7-dimethoxy-phthalide***
        3-⟨3,7-Dimethyl-octa-2,6-dienyloxy⟩-6,7-dimethoxy-phthalid
```

SUBSTANCE SEARCHING
Special Fields for Substance Name Searching

- **/CDER** Chemical name segments and property parameters for derivatives of Beilstein Registry substances.

    Chemical name segments are created by parsing at spaces and punctuation.

    Useful for: Searching for characterization derivatives of file substances, e.g.,

```
=> S DIHYDRO-FURAN-2,3-DIONE?/CN AND PHENYLHYDRAZONE/CDER,CNS
            8 DIHYDRO-FURAN-2,3-DIONE?/CN
          310 PHENYLHYDRAZONE/CDER
         3909 PHENYLHYDRAZONE/CNS
L5          5 DIHYDRO-FURAN-2,3-DIONE?/CN AND PHENYLHYDRAZONE/

=> D 1,5 HIT

L5   ANSWER 1 OF 5

CN       ***dihydro-furan-2,3-dione-3-⟨2-nitro-phenylhydrazone⟩***
         Dihydro-furan-2,3-dion-3-⟨2-nitro-phenylhydrazon⟩

L5   ANSWER 5 OF 5

CN       ***dihydro-furan-2,3-dione***
         Dihydro-furan-2,3-dion

---> Handbook Data <---
Chemical Derivative :
CDER as 2,4-dinitro-phenylhydrazone
     Reference(s):  1. Hift, Mahler, J.Biol.Chem. 198 ⟨1952⟩ 901, 909, CODEN: JBCHA3
```

SUBSTANCE SEARCHING
Special Fields for Substance Name Searching

- /PRE.EDT   Preparation Educt (Reactant)
  /PRE.RGT   Preparation Reagent
  /PRE.BPRO   Preparation By-product
  /REA.RP   Reaction Partner
  /REA.RGT   Reaction Reagent
  /REA.PRO   Reaction Product

Useful for: Searching for reaction participants, e.g.,

```
=> S (PHTHALIC (S) ANHYDRIDE) / CNS AND MESITYLENE / REA.RP
            377 PHTHALIC / CNS
           4283 ANHYDRIDE / CNS
            240 (PHTHALIC (S) ANHYDRIDE) / CNS
              5 MESITYLENE / REA.RP
L6            1 (PHTHALIC (S) ANHYDRIDE) / CNS AND MESITYLENE / REA.RP

=> D HIT

L6   ANSWER 1 OF 1

CN       ***3-chloro-phthalic acid-anhydride***
         3-Chlor-phthalsaeure-anhydrid

---> Handbook Data <---
Chemical Reaction:
REA
    Part.:  mesitylene, aluminium chloride, 1,1,2,2-tetrachloro-ethane
    Prod.:  2-chloro-6-<2,4,6-trimethyl-benzoyl>-benzoic acid,
            3-chloro-2-<2,4,6-trimethyl-benzoyl>-benzoic acid
    Reference(s):  1. Newman, Scheurer, J.Amer.Chem.Soc. 78 <1956> 5004, 5005,
         CODEN: JACSAT
```

SUBSTANCE SEARCHING   II-15
Molecular Formula Searching

- Molecular formulas are available for each Beilstein Registry substance.
- Molecular formulas are in Hill order.
- Molecular formulas are indexed for search in the /MF field and in the Basic Index.
- Molecular formulas may be searched in /MF with blanks separating the elements to avoid ambiguity resulting from combinations of element symbols that also represent elements, e.g.

    => S C23 H34 O S4 / MF   retrieves substances with 23 carbon, 34 hydrogen, 1 oxygen, and 4 sulfur atoms.

    => S C23 H34 OS4 / MF   retrieves substances with 23 carbon, 34 hydrogen, and 4 osmium atoms.

    => S C23H34OS4 / MF   retrieves both.

- For substances containing deuterium and/or tritium, only the formula with the D and/or T isotope count is indexed.

SUBSTANCE SEARCHING II-16
Element and Atom Counts

- The number of atoms of each element in a molecular formula can be searched individually as single element counts.

    EXAMPLE: 13/C
    18/H
    4/N
    3/O
    are all indexed for C13H18N4O3.

    −For molecular formulas containing halogens, an additional element count of all halogens in the formula is indexed at X.
    −For molecular formulas containing metals, an additional element count of all metals in the formula is indexed at M.

- The Element Symbol field (/ELS) contains the element symbol for each element in a molecular formula.

    EXAMPLE: C/ELS
    H/ELS
    N/ELS
    O/ELS
    are all indexed for C13H18N4O3.

- The Element Count field (/ELC) contains the total number of different elements in each molecular formula in the /MF field.

    EXAMPLE: 4/ELC is indexed for C13H18N4O3.

- The Atom Count field (/ATC) contains the total number of atoms in each molecular formula in the /MF field.

    EXAMPLE: 38/ATC is indexed for C13H18N4O3.

- The Periodic Group field (/PG) contains designations for each column in the periodic chart. A designation is generated for each element in the molecular formula. (See Appendix A for the symbol for each periodic designation.)

- The Formula Weight field (/FW) contains a formula weight calculated from the molecular formula using atomic weights listed in Appendix B.

SUBSTANCE SEARCHING
Element and Atom Counts

Search Example

QUESTION: Find information on quinoline or isoquinoline derivatives substituted only with carbon-containing substituents and having formula weights less than 150.

```
=> S (QUINOLINE OR ISOQUINOLINE) / CNS AND 1 / N AND C / ELS AND H / ELS
   AND 3 / ELC AND FW<150
          15151 QUINOLINE / CNS
           3711 ISOQUINOLINE / CNS
          87046 1 / N
         350418 C / ELS
         350117 H / ELS
           8151 6 3 / ELC
          12776 FW<150
L7           73 (QUINOLINE OR ISOQUINOLINE) / CNS AND 1 / N AND C / ELS AND
                H / ELS

=> D 15 IDE

L7  ANSWER 15 OF 73

BRN   121349  Beilstein
MF    C10 H13 N
CN    ***8-methyl-1,2,3,4-tetrahydro-quinoline***
      8-Methyl-1,2,3,4-tetrahydro-chinolin
FW    ***147.22***
SO    0-20-00-00288; 2-20-00-00185; 4-20-00-03027
LN    24301
```

SUBSTANCE SEARCHING   *II-18*
Structure Searching

- Creating query structures

    –Structure queries may be built online using the STRUCTURE command in text or graphics mode.

    –Structure queries may be built offline and uploaded using STN Express.

    –Structure queries may be built offline and uploaded using Molkick, a software package developed by the Beilstein Institute and Softron GmbH.

    –Modeling on Beilstein Registry Numbers is available within the STRUCTURE command.

    –The same definition of normalized bonds is used in BEILSTEIN and REGISTRY.

- Search types

    –SSS   Substructure search.
    –EXA   Exact structure search.
    –FAM   Family structure search.

- Search scopes

    –SAMPLE   Searches a fixed 10% sample of the file.
    –FULL   Searches the entire file.
    –RANGE   Searches a user-specified range of the file. (Beilstein Registry Numbers are used to specify ranges.)

SUBSTANCE SEARCHING II-19
Structure Searching

Search Example

QUESTION: Find information on 2,5-disubstituted tetrahydro furans represented by the structure

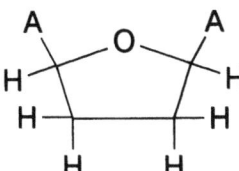

```
=> STR
:GRA R5, 1 C1, 3 C1, NOD 2 O, 6 7 A, BON ALL SE
:HCO 1 3 E1, 4 5 E2, NSP 6 7 RC, RSP I, END

L8   STRUCTURE CREATED

=> S L8 SSS SAM
SAMPLE SEARCH INITIATED 14:35:44
SAMPLE SCREEN SEARCH COMPLETED -  591 SUBSTANCES TO ITERATE
  45.3% PROCESSED    268 ITERATIONS                           6 ANSWERS
 100.0% PROCESSED    591 ITERATIONS                           9 ANSWERS
SEARCH TIME: 00.00.29

FULL FILE PROJECTIONS:  ONLINE  **COMPLETE**
                        BATCH   **COMPLETE**
PROJECTED ITERATIONS:      5187 TO      6633
PROJECTED ANSWERS:            9 TO       178

L9            9 SEA SSS SAM L8
```

SUBSTANCE SEARCHING II-20
Structure Searching

Search Example

```
=> S L8 SSS FUL
FULL SEARCH INITIATED 14:36:52
FULL SCREEN SEARCH COMPLETED -  5940 SUBSTANCES TO ITERATE
    5.7% PROCESSED    341 ITERATIONS                7 ANSWERS
   12.1% PROCESSED    720 ITERATIONS               15 ANSWERS
   20.9% PROCESSED   1243 ITERATIONS               29 ANSWERS
   42.9% PROCESSED   2549 ITERATIONS               70 ANSWERS
   71.8% PROCESSED   4266 ITERATIONS               81 ANSWERS
   89.5% PROCESSED   5317 ITERATIONS               87 ANSWERS
  100.0% PROCESSED   5940 ITERATIONS              112 ANSWERS
SEARCH TIME: 00.01.45

L10        112 SEA SSS FUL L8

=> D 1 IDE FA

L10 ANSWER 1 OF 112

BRN   380983   Beilstein
MF    C20 H16 N4 O13
CN    2,5-bis-⟨3,5-dinitro-benzoyloximethyl⟩-tetrahydro-furan
      2,5-Bis-⟨3,5-dinitro-benzoyloxymethyl⟩-tetrahydro-furan
FW    520.37
SO    4-17-00-02020
LN    17409; 10583
```

O_2N—[ring]—$C(O)OCH_2$—[furan]—$CH_2OC(O)$—[ring]—NO_2
                NO_2 NO_2

```
      Code    Field Name                          Occur.
      -------+-----------------------------------+------
      MF      Molecular Formula                    1
      CN      Chemical Name                        1
      FW      Formula Weight                       1
      SO      Beilstein Citation                   1
      LN      Lawson Number                        1
      PRE     Preparation                          1
      MP      Melting Point                        1
```

Tautomers

- Tautomer definition

$$H1-2=3 \qquad 1=2-3H$$

– A central atom (2) is connected to two heteroatoms (1 and 3).
  The central atom (2) may be C, N, P, As, Sb, S, Se, Te, Cl, Br, or I.
  The heteroatoms (1 and 3) may be N, O, S, Se, or Te.

– A double bond can be drawn between one of the heteroatoms (1 or 3) and the central atom (2).

– A single bond can be drawn between the central atom (2) and the other heteroatom (1 or 3), and the latter has either a hydrogen, a hydrogen isotope, or a charge.

Examples:

SUBSTANCE SEARCHING
General Conventions for Representing Substances

- Stereoisomers
  - Stereoisomers generally have different Beilstein Registry Numbers.

  BRN   84225   Beilstein
  CN   cis-2,5-Diphenyl-tetrahydro-furan

  BRN   91499   Beilstein
  CN   trans-2,5-Diphenyl-tetrahydro-furan

General Conventions for Representing Substances

- Stereoisomers

  –Stereoisomers with incomplete or undefined stereochemistry are collected under the same Beilstein Registry Number.

BRN 165751 Beilstein
CN   2,5-Diphenyl-tetrahydro-furan

---> Handbook Data <---
Related Structure:
RSTR
    Note(s): 1. In einem von.
RSTR
    Reference(s): 1. Tuot, Guyard, Bull.Soc.Chim.Fr. 1947 1087, 1092, CODEN: BSCFAS

    Note(s): 2. aus meso-1,4-Diphenyl-butan-1,4-diol mit Hilfe von wss. Schwefelsaeure erhaltenen opt.-inakt. Praeparat (Kp27: 210grad: D20: 1.0752; n20D: 1.5770) hat vermutlich ein Gemisch der beiden folgenden Stereoisomeren vorgelegen..

SUBSTANCE SEARCHING

General Conventions for Representing Substances

- Tautomers
    - Enol-keto tautomers are assigned the structure for which data have been reported.
    - The enol form of a substance receives a different Beilstein Registry Number than does the keto form.
    - Both the enol form and the keto form of a substance may be present in the database, each with a different Beilstein Registry Number and structure, when data are available for both forms.

BRN 384233 Beilstein

BRN 385319 Beilstein

SUBSTANCE SEARCHING
General Conventions for Representing Substances

- Tautomers

  —Tautomers fitting the CAS Registry definition of tautomers, receive different Beilstein Registry Numbers when data exist for both forms, *but* will have the same structure diagram.

  BRN 385041 Beilstein
  CN   triphenyl-pyridine-2,4-diol
       Triphenyl-pyridin-2,4-diol

SUBSTANCE SEARCHING
General Conventions for Representing Substances

- Characterization derivatives

    –Characterization derivatives receive their own Beilstein Registry Number when the structure for the derivative is unambiguously defined, the molecular formula is reported, and data are available.

    ```
    BRN   89244   Beilstein
    CN    .beta.-dihydrosantonin oxide
          .beta.-Dihydrosantoninoxid
    SO    4-19-00-01997
    NTE   stereoisomeres of unknown configuration.
    ```

    ---> Handbook Data <---
    Chemical Derivative :
    CDER oxime C15H21NO4, Mp.: 189-190 degree
        Reference(s): 1. Wedekind, Tettweiler, Chem.Ber. 64 ⟨1931⟩ 387, 394, CODEN: CHBEAM
        2. Hendrickson, Bogard, J.Chem.Soc. 1962 1678, 1687, CODEN: JCSOA9

General Conventions for Representing Substances

- Characterization derivatives

```
BRN   90570   Beilstein
CN    .beta.-dihydrosantonin oxide oxime
      .beta.-Dihydrosantoninoxid-oxim
SO    4-19-00-01997
```

---> Handbook Data <---
Melting Point:

Value (MP) (Cel)	Solv. (MP.SOL)	Ref.	Note
189.00 - 190.00		2, 3	1

Reference(s):
2. Wedekind, Tettweiler, Chem.Ber. 64 ⟨1931⟩ 387, 394, CODEN: CHBEAM
3. Hendrickson, Bogard, J.Chem.Soc. 1962 1678, 1687, CODEN: JCSOA9

Note(s):
1. Decomp.

SUBSTANCE SEARCHING
General Conventions for Representing Substances

- Characterization derivatives

  –Characterization derivatives whose structures are not unambiguously defined, are listed under the Beilstein Registry Number of the parent substance.

BRN 57533   Beilstein
CN   6'r,7'c-epoxy-5,1'-dihydroxy-7,3'-dimethyl-6',7'-dihydro-⟨2,2'⟩binaphthyl-1,4,5',8'-tetraone
     6'r,7'c-Epoxy-5,1'-dihydroxy-7,3'-dimethyl-6',7'-dihydro-⟨2,2'⟩binaphthyl-1,4,5',8'-tetraon
NTE or mirror-image.

---> Handbook Data <---
Chemical Derivative :
CDER oxime Mp.: 180-182 degree
     Reference(s): 1. Paris, Prista, Ann.Pharm.Fr. 12 ⟨1954⟩ 375, 379, CODEN: APFRAD
     2. Prista, Anais Fac. Farm. Porto 18 ⟨1958⟩ 5, 16, 36
CDER 2,4-dinitro-phenylhydrazone Mp.: 130-132 degree
     Reference(s): 1. Prista, Anais Fac. Farm. Porto 18 ⟨1958⟩ 5, 16, 36
     2. Paris, Prista, Ann.Pharm.Fr. 12 ⟨1954⟩ 375, 378, CODEN: APFRAD

SUBSTANCE SEARCHING

Using Other Substance Identifiers

- Beilstein Registry Number

```
=> DIS ACC 7809 CN FA

   ANSWER 1

CN  (±)-cis(?)-5-carboxymethyl-tetrahydro-furan-2-carboxylic acid
    (±)-cis(?)-5-Carboxymethyl-tetrahydro-furan-2-carbonsaeure
```

Code	Field Name	Occur.
MF	Molecular Formula	1
CN	Chemical Name	1
FW	Formula Weight	1
SO	Beilstein Citation	1
LN	Lawson Number	1
NTE	Notes	1
PRE	Preparation	1
MP	Melting Point	1

SUBSTANCE SEARCHING

Using Other Substance Identifiers

- Beilstein citation

    –Searched in the /SO field.
    –Number format: s-vv-ss-pppp

    s— Basic Series and Supplementary Series
       0 = Basic Series
       1 = Suppl. Series I
       2 = Suppl. Series II
       3 = Suppl. Series III
       4 = Suppl. Series IV
       5 = Suppl. Series V
    vv— Volume
    ss— Subvolume (Suppl. Series V onward)
    pppp— Page number

```
=> S 2-17-00-00013 / SO
L12           2  2-17-00-00013 / SO

=> D 2 BRN CN SO FA

L12 ANSWER 2 OF 2

BRN   79763   Beilstein
CN    (±)-propylene oxide
      (±)-Propylenoxyd
SO    0-17-00-00006; 1-17-00-00004;   ***2-17-00-00013***

      Code     Field Name                           Occur.
      -------+-----------------------------------+------
      MF       Molecular Formula                     1
      CN       Chemical Name                         1
      SY       Synonym                               1
      FW       Formula Weight                        1
      SO       Beilstein Citation                    1
      LN       Lawson Number                         1
      NTE      Notes                                 1
      PRE      Preparation                           6
      BP       Boiling Point                         2
      REA      Chemical Reaction                    21
      DEN      Density (liquid)                      1
      CTCA     Calorific Data                        1
      ORP      Optical Rotatory Power                2
```

Using Other Substance Identifiers

- CAS Registry Number^(SM)

    –CAS Registry Numbers are not yet available in the BEILSTEIN database.

    –CAS Registry Number modeling may be used to create a query in the REGISTRY File for EXACT substance match in BEILSTEIN.

```
=> FILE REGISTRY

FILE 'REGISTRY' ENTERED AT 14:48:39 ON 12 SEP 88
COPYRIGHT (C) 1988 AMERICAN CHEMICAL SOCIETY

STRUCTURE FILE UPDATES:   HIGHEST RN 116346-32-8
DICTIONARY FILE UPDATES:  10 SEPT 88 (880910 / ED)   HIGHEST RN 116323-40-1

=> STR 78160-12-0
:END

L1   STRUCTURE CREATED
```

SUBSTANCE SEARCHING
Using Other Substance Identifiers

- CAS Registry Number

```
=> FILE BEILSTEIN

FILE 'BEILSTEIN' ENTERED AT 14:49:18 ON 12 SEP 88
COPYRIGHT (c) 1988 Springer-Verlag

=> S L1 SSS S EXA

SAMPLE SEARCH INITIATED 14:49:51
SAMPLE SCREEN SEARCH COMPLETED -     1 SUBSTANCES TO ITERATE
100.0% PROCESSED    1 ITERATIONS                          1 ANSWERS
SEARCH TIME: 00.00.02

FULL FILE PROJECTIONS:  ONLINE   **COMPLETE**
                        BATCH    **COMPLETE**
PROJECTED ITERATIONS:          1 TO        39
PROJECTED ANSWERS:             1 TO        39

L2               1 SEA EXA SAM L1
```

SUBSTANCE SEARCHING
Using Other Substance Identifiers

- CAS Registry Number

```
=> D 1 IDE FA

L2   ANSWER 1 OF 1

BRN  192313  Beilstein
MF   C14 H11 N O S
CN   2-amino-7-methyl-thioxanthen-9-one
     2-Amino-7-methyl-thioxanthen-9-on
FW   241.31
SO   4-18-00-07968; 1-18-00-00574
LN   20604
```

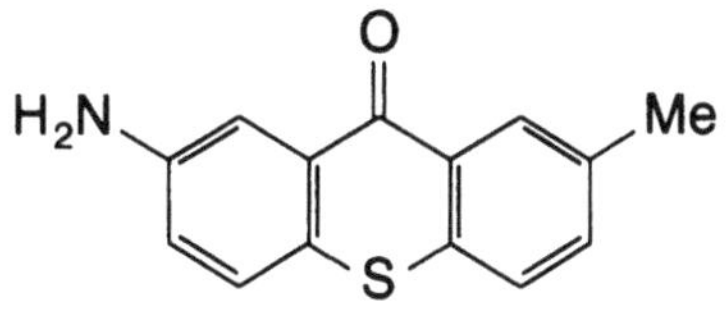

[Structure: 2-amino-7-methyl-thioxanthen-9-one]

```
Code    Field Name                      Occur.
--------+-------------------------------+------
MF      Molecular Formula               1
CN      Chemical Name                   1
FW      Formula Weight                  1
SO      Beilstein Citation              1
LN      Lawson Number                   1
PRE     Preparation                     2
MP      Melting Point                   2
CPD     Crystal Property Description    2
```

SUBSTANCE SEARCHING   *II-34*
Similarity Searching

- The Lawson Number (/LN) may be used to search for substances with structures similar to a substance already retrieved from the database, e.g., positional isomers, steroisomers, etc.

- Lawson Numbers represent structure fragments created by segmenting the molecule. In simple terms, segmentation occurs at all chain heteroatoms having at least one attached carbon atom; e.g.,

has the following Lawson Numbers:

Fragment	Lawson Number
H₂N-pyridine-H₂N	27515
HO-phenyl-(NH₂)(NH₂)	14893
Me—OH	289

- Every substance in the file has at least one Lawson Number, the average being 2–3 Lawson Numbers per substance.

- Usually, the smaller the Lawson Number, the more common the fragment, so it is best to use the largest Lawson Number for similarity searching.

- Lawson Number searching is not substructure searching or Markush searching. It retrieves similar structures, some of which may be considered to be "false drops".

Search Example

QUESTION 1: Are other positional isomers of the substance shown in the following answer in the file?

BRN   195896   Beilstein

LN   20604

```
=> S C14 H11 N O S/MF AND 20604/LN
          63 C14 H11 N O S/MF
          53 20604/LN
L4         4 C14 H11 N O S/MF AND 20604/LN

=> D 1-4 STR LN

L4   ANSWER 1 OF 4

BRN   195896   Beilstein
```

LN   20604

Search Example

L4 ANSWER 2 OF 4

BRN 193405 Beilstein

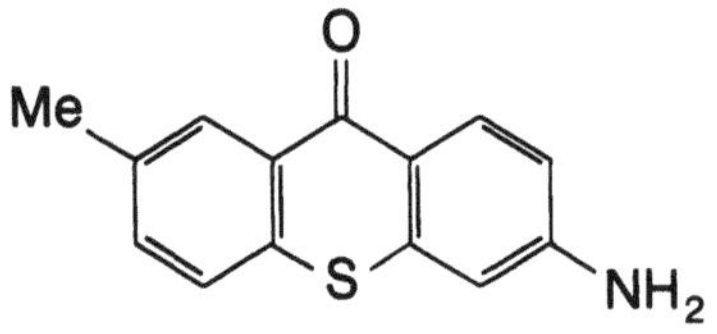

LN 20604

L4 ANSWER 3 OF 4

BRN 192313 Beilstein

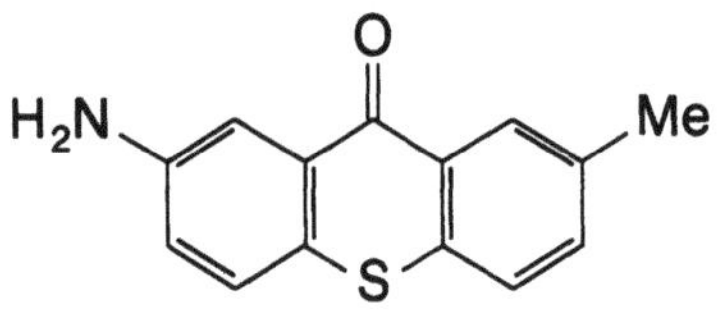

LN 20604

L4 ANSWER 4 OF 4

BRN 16125 Beilstein

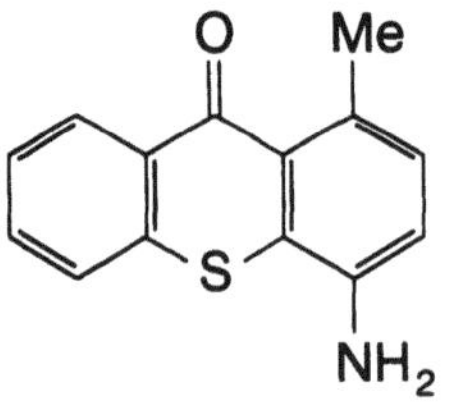

LN 20604

SUBSTANCE SEARCHING *II-37*
Displaying Substance Information

- The IDE format may be used to display all substance information.
- Fields displayed in the IDE format are:

BRN	Beilstein Registry Number
MF	Molecular Formula
LSF	Linearized Structure Formula
CN	Chemical Name
SY	Synonym
FW, MW	Formula Weight
SO	Beilstein Citation
NTE	Note
LN	Lawson Number
STR	Structure

- All field codes listed above, except NTE, may be used in DISPLAY or PRINT.
- Hit term highlighting is available in the fields listed above except for NTE and STR.
- Use HELP DIDE online to get a list of the substance display fields.

III. Searching for Physical Property Information

Types of Property Information Available
Searching for a Property
Searching for a Specific Property Value or Range of Values
Displaying Property Information

SEARCHING FOR PHYSICAL PROPERTY INFORMATION
Types of Property Information Available

- Data on electrochemical behavior
    - Isoelectric points
    - Dissociation exponents
    - Redox potentials
    - Etc.
- Data on electrical behavior
    - Dielectric constants
    - Dielectric static constants
- Magnetic data
    - Magnetic susceptibility
- Data on mechanical properties
    - Density
    - Linear expansion coefficients
    - Molar volume
    - Surface tension
- Optical data
    - Circular dichroism
    - Mutarotation
    - Optical rotation dispersion
    - Optical rotatory power
    - Refractive index
- State of aggregation data
    - Crystal lattice parameters
    - Melting point
    - Boiling point
    - Vapor pressure
    - Critical temperatures and pressures
    - Critical densities
    - Etc.
- Structure and energy parameters
    - Nuclear quadrupole coupling constants
    - Bond moments
    - Dipole moments
    - Energy of dissociation
    - Rotational constants
    - Ionization potentials
    - Etc.
- Spectral data
    - Fluorescence maxima/spectra
    - Phosphorescence maxima/spectra
    - Electronic absorption maxima/spectra
    - Infrared maxima/spectra
    - Raman maxima/spectra
    - NMR absorption/spectra
    - Mass spectra
    - Etc.
- Thermodynamic data
    - Enthalpy of formation and combustion
    - Entropies
    - Heat capacities
    - Etc.
- Transport phenomena
    - Dynamic viscosity
    - Kinematic viscosity
    - Etc.

SEARCHING FOR PHYSICAL PROPERTY INFORMATION

General Information of Property Searching

- Physical properties may be present in the file as specific data values or may be indicated by pointers to the original literature, called controlled terms or keywords.

```
BRN    385386   Beilstein


--->  Handbook Data  <---
Melting Point:
Value (MP)              Solv. (MP.SOL)              Ref.    Note
(Cel)
-----------------------+---------------------------+-------+------
249.00 - 254.00         ethanol                      2       1

Reference(s):
2. Yamane, Nippon Kagaku Zasshi 80 〈1959〉 534, CODEN: NPKZAZ  CA: 1961 4500

Note(s):
1. Decomp.


BRN    333502   Beilstein

--->  Handbook Data  <---
CTCA Calorific Data: Cryoscopic constant
     Reference(s):  1. Hantzsch, Z.Phys.Chem.Stoechiom.Verwandtschaftsl. 61
           〈1908〉,281, CODEN: ZEPCAC
     Note(s):2. Kryoskopisches Verhalten in Schwefelsaeuremonohydrat.
```

SEARCHING FOR PHYSICAL PROPERTY INFORMATION
General Information of Property Searching

- Controlled terms for single component substances are posted in the /CT search field.
- Controlled terms for multicomponent substances are posted in the /CTM search field.
- The controlled words or keywords are hierarchical, i.e., specific controlled terms are upposted to broader controlled terms. (See Appendix C for a list of the controlled terms.)
- The controlled terms can be used to search for specific kinds of data but not for exact numeric values.

```
=> E CALORIFIC DATA/CT
E1         114 BASICITY/CT
E2           1 BOND ENERGY/CT
E3         346 CALORIFIC DATA/CT
E4           1 COLE-COLE DIAGRAM/CT
E5           1 COLLISION CROSS-SECTION/CT
E6         295 CONFORMATION/CT
E7           2 CONFORMER EQUILIBRIUM/CT
E8         407 COUPLING PHENOMENA/CT
E9           6 CRITICAL FREQUENCY (OR WAVELENGTH)/CT
E10        131 CRYOSCOPIC CONSTANT/CT
E11          3 CRYSTAL GROWTH/CT
E12         16 CRYSTAL HABIT/CT

=> S E10 AND E3
           131 "CRYOSCOPIC CONSTANT"/CT
           346 "CALORIFIC DATA"/CT
L1         131 "CRYOSCOPIC CONSTANT"/CT AND "CALORIFIC DATA"/CT

=> E SOLIDIFICATION/CTM
E1          24 PRESSURE-SURFACE ISOTHERM/CTM
E2           1 RATE OF DISSOLUTION/CTM
E3           0 SOLIDIFICATION/CTM
E4          47 SOLIDIFICATION DIAGRAM/CTM
E5          19 SOLIDIFICATION POINTS OF MIXTURES/CTM
E6          10 SOLUBILIZING/CTM
E7          85 SOLUTION BEHAVIOUR/CTM
E8          34 SOLUTION EQUILIBRIUM/CTM
```

SEARCHING FOR PHYSICAL PROPERTY INFORMATION

General Information of Property Searching

- The property hierarchy field (/PH) contains the names and field codes for all fields containing property data, plus the upposted controlled terms.
- The /PH field can be used to search for the presence of a property, regardless of whether a specific value is recorded in the file or the information is recorded as a controlled term.

```
=> E BP/PH
E1           1 BOND MOMENT/PH
E2         148 BOUNDARY SURFACE PHENOMENA/PH
E3       45898 BP/PH
E4         346 CALORIFIC DATA/PH
E5        2381 CDER/PH
E6          48 CDIC/PH
E7        2381 CHEMICAL DERIVATIVE /PH
E8       38231 CHEMICAL REACTION/PH
E9          48 CIRCULAR DICHROISM/PH
E10        536 CLP/PH
E11          1 COLE-COLE DIAGRAM/PH
E12          1 COLLISION CROSS-SECTION/PH
```

SEARCHING FOR PHYSICAL PROPERTY INFORMATION
Searching for a Property

Search Example

QUESTION: Find porphyrins for which infrared data has been reported. Have NMR spectra been reported for any of these substances?

```
=> E INFRARED/PH 7
E1              84 HVAP/PH
E2              29 IEP/PH
E3               0 INFRARED/PH
E4            4579 INFRARED MAXIMUM/PH
E5            5825 INFRARED SPECTRUM/PH
E6            3550 INP/PH
E7              40 INTENSITY OF IR BANDS/PH

=> S (E4 OR E5) AND PORPHYRIN?/CNS
              4579 "INFRARED MAXIMUM"/PH
              5825 "INFRARED SPECTRUM"/PH
               628 PORPHYRIN?/CNS
L7              41 ("INFRARED MAXIMUM"/PH OR "INFRARED SPECTRUM"/PH) AND

=> D 3 CN FA

L7   ANSWER 3 OF 41
CN   ***3,3',3'',3'''-⟨3,8,13,17-tetramethyl-porphyrin-2,7,12,18-tetrayl⟩-tetra-
     propionic acid tetramethyl ester***
     3,3',3'',3'''-⟨3,8,13,17-Tetramethyl-porphyrin-2,7,12,18-tetrayl⟩-tetra-
     propionsaeure-tetramethylester

     Code    Field Name                              Occur.
     -------+--------------------------------------+------
     MF      Molecular Formula                        1
     CN      Chemical Name                            1
     FW      Formula Weight                           1
     SO      Beilstein Citation                       1
     LN      Lawson Number                            1
     PRE     Preparation                              2
     MP      Melting Point                           12
     INP     Isolation from Natural Product           2
     CPD     Crystal Property Description            11
     CTCR    Crystal Phase                            1
     NMRS    NMR Spectrum                             1
     IRS     Infrared Spectrum                        2
     EAS     Electronic Absorption Spectrum           4
     CTLS    Liquid/Solid Systems                     1
```

Search Example

```
=> S L7 AND NMRS/FA
             370 NMRS/FA
L8             4 L7 AND NMRS/FA

=> D CN IRS NMRS

L8   ANSWER 1 OF 4
CN      ***3,3',3'',3'''-⟨3,8,13,17-tetramethyl-porphyrin-2,7,12,18-tetrayl⟩-tetra-propionic
        acid tetramethyl ester***
        3,3',3'',3'''-⟨3,8,13,17-Tetramethyl-porphyrin-2,7,12,18-tetrayl⟩-tetra-propionsaeure-
        tetramethylester

---> Handbook Data <---
Infrared Spectrum:
IRS   3500 - 650 cm**-1
   Solv:  nujol
   Reference(s):  1. Falk,Willis, Austral.J.Sci.Res.Ser.A 4 ⟨1951⟩ 579,584, CODEN:
                  AJSRA2
IRS   1300 - 700 cm**-1
   Solv:  nujol
   Reference(s):  1. Gray et al., Biochem.J. 47 ⟨1950⟩ 87,90, CODEN: BIJOAK

---> Handbook Data <---
NMR Spectrum:
NMRS
   Nucl:  1H
   Solv:  CDCl3
   Reference(s):  1. Becker,Bradley, J.Chem.Phys. 31 ⟨1959⟩ 1413, CODEN: JCPSA6
                  2. Becker et al., J.Amer.Chem.Soc. 83 ⟨1961⟩ 3743,3745,3747, CODEN: JACSAT
```

SEARCHING FOR PHYSICAL PROPERTY INFORMATION
Searching for a Specific Data Value or Values

- Property values are posted as numeric search fields which can be searched with the following numeric operators:

$$>, <, >=, <=, =$$

- Property values may be input using scientific notation, e.g., 12,345 or 1.2345E4.

```
= > S MP = 100
L11           2353 MP = 100

= > D 50,100 HIT

L11 ANSWER 50 OF 2353

---> Handbook Data <---
Melting Point:
Value (MP)              Solv. (MP.SOL)           Ref.   Note
(Cel)
- ----------------------+------------------------+-----+-----
100.00                  ethanol                  1

Reference(s):
1. Hesse, J.Prakt.Chem. ⟨2⟩ 57 ⟨1898⟩, 315, CODEN: JPCEAO

L11 ANSWER 100 OF 2353

---> Handbook Data <---
Melting Point:
Value (MP)              Solv. (MP.SOL)           Ref.   Note
(Cel)
- ----------------------+------------------------+-----+-----
99.00 - 101.00          aq. methanol             1

Reference(s):
1. Chatterjea, J.Indian Chem.Soc. 30 ⟨1953⟩ 1, 8, CODEN: JICSAH
```

SEARCHING FOR PHYSICAL PROPERTY INFORMATION
Searching for a Specific Data Value or Values

- Use the HELP commands online or consult the STN Database Summary Sheet for the appropriate search field(s) to use to search for a specific property.

```
=> HELP SPHYS
There are more than 100 search fields in the Beilstein file that contain values for Physical
Data of chemical substances. More information about the search fields which can be
used to search for specific properties is available by typing 'HELP' and one of the
following keywords at an arrow prompt, e.g., => HELP SELE.
```

Keyword	Topic
SECB	Electrochemical Behaviour
SELE	Electrical Data
SMAG	Magnetic Data
SMCS	Multi-Component System Data
SMEC	Mechanical Properties
SOPT	Optical Properties
SSAG	State of Aggregation
SSEP	Structure and Energy Parameters
SSPE	Spectral Data
STHE	Thermodynamic Data
STRA	Transport Phenomena

For information on property names and how to search them type 'HELP PROPERTIES' at an arrow prompt (=>).

SEARCHING FOR PHYSICAL PROPERTY INFORMATION
Searching for a Specific Data Value or Values

- Each property search field has its own specific units.
- Use HELP online or consult the Database Summary Sheet for the appropriate units.

```
=> HELP SSAG
The search fields that can be used for information about the State of Aggregation of
substances in the Beilstein File are

Definition                        Search Code      Units
Boiling Point:                    /BP              Cel
  Pressure                        /BP.P            Torr
Critical Density                  /CRD             g/cm**3
Critical Pressure                 /CRP             Torr
Critical Temperature              /CRT             Cel
Critical Volume                   /CRV             cm**3/mol
Crystal Lattice Parameter:        /CLP             Angstroem
  Angle                           /CLP.ANG         deg
  Number of Formula Units         /CLP.NFU         (none)
Crystal Transition Point          /CTP             Cel
Crystal Property Description*     /CPD             (none)
Crystal Space Group               /CSG, /SG        (none)
Crystal System                    /CSYS            (none)
Decomposition Point:              /DP              Cel
  Solvent                         /DP.SOL          (none)
Liquid Transition Point           /LTP             Cel
Melting Point:                    /MP              Cel
  Solvent                         /MP.SOL          (none)
Sublimation Point:                /SP              Cel
  Pressure                        /SP.P            Torr
Triple Point                      /TP              Cel
Vapour Pressure:                  /VP              Torr
  Temperature                     /VP.T            Cel

* Note: input in German
All fields are numeric except /CPD /CSG, /SG, /CSYS, /DP.SOL, and /MP.SOL, which
are text fields. Numeric fields can be searched with numeric operators (<, >, <=, >=, =,
-). Enter 'HELP SNUMERIC' at an arrow prompt for information about searching numeric
ranges.
```

SEARCHING FOR PHYSICAL PROPERTY INFORMATION
Searching for a Specific Data Value or Values

- Many properties have parameters associated with them.
- Parameters are in separate search fields whose field code is in the format

    /property field code.parameter code

- Use (P) proximity to associate a property with a parameter.

```
=> S 1.2750-1.2755/RI (P) 589/RI.W (P) 20/RI.T
            3  1.2750-1.2755/RI
        17433  589/RI.W
         8393  20/RI.T
L12         1  1.2750-1.2755/RI (P) 589/RI.W (P) 20/RI.T

=> D

L12 ANSWER 1 OF 1

BRN   209701   Beilstein
MF    C5 F11 N
CN    undecafluoro-piperidine
      Undecafluor-piperidin
FW    283.04
SO    4-20-00-01380
LN    24083
```

```
            F
            |
       F    N    F
      F         F
       F         F
        F   F
         F F
```

```
---> Handbook Data <---
Refractive Index:
Value (RI)        Wavel. (RI.W)      Temp. (RI.T)      Ref.     Note
(--)              (nm)               (Cel)
----------------+------------------+-----------------+--------+------
1.27520           589.00             20.0              1

Reference(s):
1. Simmons et al., J.Amer.Chem.Soc. 79 ⟨1957⟩ 3429,3430, CODEN: JACSAT
```

Search Example

QUESTION: What substances containing the following structure fragment and at least two other Cl atoms have boiling points above 240°C at 760 Torr pressure?

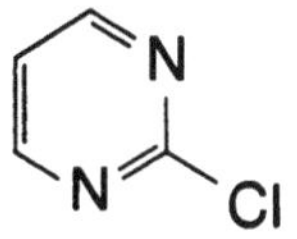

```
=> STR
:GRA R6, 2 C1, NOD 1 3 N, 7 CL, BON R 1 2 N, 2 7 SE, END
L5   STRUCTURE CREATED

=> S L5 FULL
FULL SEARCH INITIATED 19:45:26
FULL SCREEN SEARCH COMPLETED -   626 SUBSTANCES TO ITERATE
  45.7% PROCESSED      286 ITERATIONS                          282 ANSWERS
 100.0% PROCESSED      626 ITERATIONS                          594 ANSWERS
SEARCH TIME: 00.00.30

L6         594 SEA SSS FUL L5

=> S L6 AND CL>2
         3149 CL>2
L7         48 L6 AND CL>2
```

SEARCHING FOR PHYSICAL PROPERTY INFORMATION III-13
Searching for a Specific Data Value or Values

Search Example

```
= > S L7 AND BP>240 (P) 760/BP.P
        3340 BP>240
        4249 760/BP.P
        1521 BP>240 (P) 760/BP.P
L83 L7 AND BP>240 (P) 760/BP.P

= > D

L8 ANSWER 1 OF 3

BRN    155912   Beilstein
MF     C5 H Cl5 N2
CN     ***2,4-dichloro-6-trichloromethyl-pyrimidine***
       2,4-Dichlor-6-trichlormethyl-pyrimidin
FW     266.34
SO     0-23-00-00093
LN     28062
```

```
---> Handbook Data <---
Boiling Point:
Value (BP)                    Press. (BP.P)              Ref.    Note
(Cel)                         (Torr)
--------------------------+-------------------------+------+------
264.00 - 266.00               760                        1

Reference(s):
1. Gabriel, Colman, Chem.Ber. 35 ⟨1902⟩, 1569, CODEN: CHBEAM
```

SEARCHING FOR PHYSICAL PROPERTY INFORMATION
Displaying Property Data

- The HIT format displays all fields containing the hit terms.
- For fields containing multiple entries, only the entry matching the query is displayed.

```
=> S 1.2750/PI
L15            2 1.2750/RI

=> D 2 HIT

L15 ANSWER 2 OF 2

---> Handbook Data <---
Refractive Index:
Value (RI)          Wavel. (RI.W)        Temp. (RI.T)         Ref.    Note
(--)                (nm)                 (Cel)
----------------+----------------+----------------+-----+-----
1.14037             667.80               15.0                 1
- 1.43951           - 434.10             - 15.0

Reference(s):
1. Timmermans,Hennaut-Roland, J.Chim.Phys.Phys.Chim.Biol. 56 ⟨1959⟩ 984,1008,
   CODEN: JCPBAN
```

```
=> D 2 RI

L15 ANSWER 2 OF 2
```

---> Handbook Data <---
Refractive Index:

Value (RI) (--)	Wavel. (RI.W) (nm)	Temp. (RI.T) (Cel)	Ref.	Note
1.42385	589.00	15.0	1	
1.42140	589.00	20.0	2	
1.42161	589.00	20.0	3	
1.41871	589.00	25.0	3	
1.41562	589.00	30.0	3	
1.14037	667.80	15.0	1	
- 1.43951	- 434.10	- 15.0		
1.41822	656.30	20.0	2	
- 1.43619	- 435.80	- 20.0		
1.41880	643.80	20.0	4	
- 1.43630	- 435.90	- 20.0		
1.42867	495.70	20.0	5, 6	
- 1.56206	- 248.40	- 20.0		
1.40703	656.30	21.6	7	
1.42470	434.00	21.6	7	
1.42170	589.00	19.3	8	
1.41840	656.30	20.0	9	
1.42200	589.00	20.0	9	
1.42970	486.10	20.0	9	
1.43670	434.00	20.0	9	

Reference(s):
1. Timmermans,Hennaut-Roland, J.Chim.Phys.Phys.Chim.Biol. 56 ⟨1959⟩ 984,1008, CODEN: JCPBAN
2. Guthrie et al., J.Amer.Chem.Soc. 74 ⟨1952⟩ 4662,4663, CODEN: JACSAT
3. Kobe et al., J.Chem.Eng.Data 1 ⟨1956⟩ 50,52, CODEN: JCEAAX
4. Hughes,Johnson, J.Amer.Chem.Soc. 53 ⟨1931⟩ 737,745, CODEN: JACSAT
5. Le Fevre et al., J.Chem.Soc. 1959 1188,1189, CODEN: JCSOA9
6. Holbro, Helv.Phys.Acta 10 ⟨1937⟩ 431,449, CODEN: HPACAK
7. Nasini, Carrara, Gazz.Chim.Ital. 24 I ⟨1894⟩, 278, CODEN: GCITA9
8. Landrieu, Baylocq, Johnson, Bull.Soc.Chim.Fr. ⟨4⟩ 45 ⟨1929⟩, 43, CODEN: BSCFAS
9. v. Auwers, Liebigs Ann.Chem. 408 ⟨1915⟩, 270, CODEN: LACHDL

SEARCHING FOR PHYSICAL PROPERTY INFORMATION
Displaying Property Data

- The default format displays the substance identification information, *plus* the entries matching the query.

- The following general formats are available for displaying various classes of property information. (See the Database Summary Sheet for all property display codes.)

Display Format	Content
PHYS	All physical properties: includes ECB, ELE, MAG, MCS, MEC, OPT, SAG, SEP, SPE, THE, TRA
ECB	Electrochemical behavior
ELE	Electrical data
MAG	Magnetic data
MCS	Multicomponent system data
MEC	Mechanical properties
OPT	Optical data
SAG	State of aggregation data
SEP	Structure and energy parameters
SPE	Spectral data
THE	Thermodynamic data
TRA	Transport phenomena

- Use DIS FA to see the display fields associated with an answer.

IV. Searching for Preparations and Reactions

General Information on Preparations and Reactions
Searching for Preparations
Searching for Reactions
Displaying Preparation and Reaction Information

SEARCHING FOR REACTIONS AND PREPARATIONS   IV-2
General Information

- Preparations

    −All reported preparations were indexed in the early part of the handbook.
    −Current selection criteria are

    > Yield
    > Ease of handling of educts
    > Number of steps

- Information on isolation of substances from natural products is included in the database.

- Reactions

    −Reactions judged to be new and/or interesting are included in the database.

SEARCHING FOR PREPARATIONS AND REACTIONS
Searching for Preparations

- Preparative information about Beilstein Registry substances can be found in the following search fields:

/PRE.EDT	Educt(s) used in the reaction
/PRE.RGT	Reagent(s) used in the reaction
/PRE.BPRO	By-product(s) of the reaction

- Preparation information, including substance names, is parsed at spaces and punctuation and posted in the appropriate /PRE search field.
- Terms from the /PRE search fields are also in the Basic Index.
- Use (P) proximity when searching for terms from the same preparation.

```
=> S NAPHTHYLAMINE/PRE.EDT (P) PEROXIDE/PRE.RGT
           950 NAPHTHYLAMINE/PRE.EDT
          2172 PEROXIDE/PRE.RGT
L28         14 NAPHTHYLAMINE/PRE.EDT (P) PEROXIDE/PRE.RGT

=> D PRE

L28 ANSWER 1 OF 14

---> Handbook Data <---
Preparation:
PRE
    Educt:  2-⟨4-bromo-phenyl⟩-⟨naphtho-1'.2':4.5-triazole⟩
    Reag:   alkaline permanganate
    Reference(s):  1. Charrier, Gazz.Chim.Ital. 53 ⟨1923⟩, 842, CODEN: GCITA9
           54 ⟨1924⟩,655, GCITA9
PRE
    Educt:  ⟨4-bromo-benzen⟩-⟨1 azo 1⟩-naphthylamine-(2)
    Reag:   hydrogen peroxide, glacial acetic acid
    Reference(s):  1. Charrier,Crippa, Gazz.Chim.Ital. 55 ⟨1925⟩, 26, CODEN: GCITA9
```

- Other fields which contain prepared substances or preparative information are

/REA.PRO	Reaction product
/INP	Isolation from natural products

SEARCHING FOR PREPARATIONS AND REACTIONS
Searching for Preparations

Search Example

QUESTION: Find ways of preparing substances with the following substructure:

```
=> STR
:GRA R66, 10 C1, NOD 7 O, 11 PH, BON ALL SE, R 1 2 N, RSP I, END
L16   STRUCTURE CREATED

=> S L16 FULL
FULL SEARCH INITIATED 20:20:37
FULL SCREEN SEARCH COMPLETED -   168 SUBSTANCES TO ITERATE
100.0% PROCESSED     168 ITERATIONS                              71 ANSWERS
SEARCH TIME: 00.00.17

L17             71 SEA SSS FUL L16

=> S L17 AND (PRE OR INP)/FA
         325747 PRE/FA
           3550 INP/FA
L18             64 L17 AND (PRE OR INP)/FA
```

Searching for Preparations

Search Example

```
= > D CN STR FA

L18 ANSWER 1 OF 64

CN   4-Acetoxy-5,7-dimethoxy-3-〈4-methoxy-phenyl〉-2,4-diphenyl-chroman
```

Code	Field Name	Occur.
MF	Molecular Formula	1
CN	Chemical Name	1
FW	Formula Weight	1
SO	Beilstein Citation	1
LN	Lawson Number	1
PRE	Preparation	1
MP	Melting Point	1

Search Example

```
= > S (PHENYL (W) CHROMAN?)/REA.PRO
         5618 PHENYL/REA.PRO
          259 CHROMAN?/REA.PRO
L19        70 (PHENYL (W) CHROMAN?)/REA.PRO

= > D 1-3 HIT

L19 ANSWER 1 OF 70

--- > Handbook Data < - --
Chemical Reaction:
REA
    Part.:  benzene, ether, phenylmagnesium bromide
    Prod.:  2.2-diphenyl-⟨1.2-chromene⟩, 2-oxy-2.4-di-phenyl-chroman
    Reference(s):  1. Loewenbein, Pongracz, Spiess, Chem.Ber. 57 ⟨1924⟩, 1519,
        1524, CODEN: CHBEAM

L19 ANSWER 2 OF 70

--- > Handbook Data < - --
Chemical Reaction:
REA
    Part.:  platinum, acetic acid
    Detail: Hydration
    Prod.:  2,3,5,7-tetraacetoxy-2-⟨3,4-diacetoxy-phenyl⟩-chroman,
            2-acetoxy-1-⟨3,4-diacetoxy-phenyl⟩-3-⟨2,4,6-triacetoxy-phenyl⟩-
            propan-1-ol,
            2-acetoxy-1-⟨3,4-diacetoxy-phenyl⟩-3-⟨2,4,6-triacetoxy-phenyl⟩-propane-1
    Reference(s):  1. Freudenberg et al., Liebigs Ann.Chem. 518 ⟨1935⟩ 37, 56,
        CODEN: LACHDL
    2. Weinges, Chem.Ber. 94 ⟨1961⟩ 3032, 3040, CODEN: CHBEAM

L19 ANSWER 3 OF 70

--- > Handbook Data < - --
Chemical Reaction:
REA
    Part.:  zinc dust, acetic acid anhydride
    Temp:   100.0 Cel
    Detail: Erhitzen des Reaktionsprodukts mit waessrig-alkoholischer Schwefelsaeure
    Prod.:  3-oxy-5.7-dimethoxy-4-⟨3.4-dimethoxy-phenyl⟩-chroman
    Reference(s):  1. Nierenstein, J.Chem.Soc. 119 ⟨1921⟩, 167, CODEN: JCSOA9
```

Searching for Reactions

- Reaction information for Beilstein Registry substances can be found in the following search fields:

    /PRE.EDT   Educt(s) used in the reaction to prepare Registry substances
    /PRE.RGT   Reagent(s) used in the reaction to prepare Registry substances
    /PRE.BPRO   By-product(s) of the reaction to prepare Registry substances
    /REA.RP   Reaction partner in reactions of Registry substances
    /REA.RGT   Reagent(s) used in reactions of Registry substances
    /REA.PRO   Products of reactions of Registry substances

- Reaction information, including substance names, is parsed at spaces and punctuation and posted in the appropriate PRE or REA search fields.

- Terms from the PRE and REA search fields are also in the Basic Index.

- Substance names in the PRE and REA search fields are in English.

- Use (P) proximity when searching for terms from the same preparation or reaction.

- Use (S) proximity to require that name segments come from the same name.

SEARCHING FOR PREPARATIONS AND REACTIONS
Searching for Reactions of the Type A + B

Search Example

QUESTION: Find reactions that use furfural and malonic acid as reactants.

```
=> S FURFURAL/CN AND (MALONIC (W) ACID)/REA.RP
             1 FURFURAL/CN
           113 MALONIC/REA.RP
         17138 ACID/REA.RP
           113 (MALONIC(W)ACID)/REA.RP
L1           1 FURFURAL/CN AND (MALONIC(W)ACID)/REA.RP

=> S MALONIC ACID/CN AND FURFURAL/REA.RP
             0 MALONIC ACID/CN
             7 FURFURAL/REA.RP
L2           0 MALONIC ACID/CN AND FURFURAL/REA.RP

=> S (MALONIC (W) ACID ) (P) FURFURAL
          4119 MALONIC
        190528 ACID
          1185 FURFURAL
L3          14 (MALONIC (W) ACID ) (P) FURFURAL

=> S L1 OR L3

L5          15 L1 OR L3
```

Search Example

```
=> D 1,7 HIT

L5    ANSWER 1 OF 15

---> Handbook Data <---
Preparation:
PRE
    Educt:  furfural, malonic acid dimethyl ester
    Reag:   piperidine, acetic acid, benzene
    Reference(s):  1. Beyler, Sarett, J.Amer.Chem.Soc. 74 ⟨1952⟩ 1397, 1399,
        CODEN: JACSAT

L5    ANSWER 7 OF 15

CN    Furfural

---> Handbook Data <---
Chemical Reaction:

REA
    Part.:  malonic acid, pentylamine
    Prod.:  furfurylidenemalonic acid
    Reference(s):  1. Pat. No.: 164296, D.R.P., Knoevenagel

REA
    Part.:  malonic acid, glacial acetic acid
    Prod.:  furfurylidenemalonic acid
    Reference(s):  1. Liebermann, Chem.Ber. 27 ⟨1894⟩, 285, CODEN: CHBEAM

REA
    Part.:  malonic acid, alcoholic ammonia (2 mol)
    Prod.:  furfurylidenemalonic acid
    Reference(s):  1. Knoevenagel, Chem.Ber. 31 ⟨1898⟩, 2614, CODEN: CHBEAM

REA
    Part.:  malonic acid, alcoholic ammonia (1 mol)
    Prod.:  .beta.-⟨.alpha.-furyl⟩-acrylic acid
    Reference(s):  1. Knoevenagel, Chem.Ber. 31 ⟨1898⟩, 2614, CODEN: CHBEAM

REA
    Part.:  malonic acid diethyl ester, acetic acid anhydride
    Prod.:  furfurylidenemalonic acid diethyl ester
    Reference(s):  1. Marckwald, Chem.Ber. 21 ⟨1888⟩, 1081, CODEN: CHBEAM
```

SEARCHING FOR REACTIONS AND PREPARATIONS
Searching for Reactions of the Type A → B

Search Example

QUESTION: Find ways of preparing pyridoxal from pyridoxine.

```
=> S (PYRIDOXAL/CN AND PYRIDOXINE/PRE.EDT) OR (PYRIDOXINE/CN AND
PYRIDOXAL/REA.PRO)
            1 PYRIDOXAL/CN
           26 PYRIDOXINE/PRE.EDT
            1 PYRIDOXINE/CN
            6 PYRIDOXAL/REA.PRO
L6          2 (PYRIDOXAL/CN AND PYRIDOXINE/PRE.EDT) OR PYRIDOX/

=> D 1-2 HIT

L6   ANSWER 1 OF 2

CN   Pyridoxal

---> Handbook Data <---
Preparation:
PRE
     Educt:   pyridoxine
     Reag:    manganese (IV)-oxide, water, sulfuric acid
     Reference(s):   1. Viscontini et al., Helv.Chim.Acta 34 〈1951〉 1834,1839,
          CODEN: HCACAV
     2. Heyl, J.Amer.Chem.Soc. 70 〈1948〉 3434, CODEN: JACSAT

L6   ANSWER 2 OF 2

CN   ***pyridoxine***
     Pyridoxin

---> Handbook Data <---
Chemical Reaction:
REA
     Part.:   potassium permanganate, HCl
     Detail:  in wss. Loesungen
     Prod.:   pyridoxal
     Reference(s):   1. Harris et al., J.Amer.Chem.Soc. 66 〈1944〉 2088, 2090,
          CODEN: JACSAT
REA
     Part.:   manganese (IV)-oxide, H2SO4
     Detail:  in wss. Loesungen
     Prod.:   pyridoxal
     Reference(s):   1. Viscontini et al., Helv.Chim.Acta 34 〈1951〉 1834, 1838,
          CODEN: HCACAV
     2. Heyl, J.Amer.Chem.Soc. 70 〈1948〉 3434, CODEN: JACSAT
```

SEARCHING FOR PREPARATIONS AND REACTIONS
Displaying Reaction Information

- The HIT format displays only the reactions or preparations that contain the hit terms.
- The default format displays the substance identification information, *plus* the reactions or preparations matching the query.
- The following formats are available for displaying reaction information:

Display Format	Content
CHE	All chemical data (includes ISO, REA, and SYN)
ISO	Isolation information
REA	Reaction data
SYN	Synthesis data (preparation, isolation, and purification)

V. Other Search Information

Additional Search Hints
Searching for References
System Limits
Help

OTHER SEARCH INFORMATION
Additional Search Hints

- Additional text fields are available for search.

```
=> S ARTEMISIA / INP
L3            22 ARTEMISIA / INP

=> d 1-2 cn hit

L3   ANSWER 1 OF 22

CN    6-hydroxy-3-oxo-eudesma-1,4-dien-12-oic acid-lactone
      6-Hydroxy-3-oxo-eudesma-1,4-dien-12-saeure-lacton

---> Handbook Data <---
Isolation from Natural Product:
INP   in den Bluetenkoepfchen von Artemisia mexicana Willd., A. neomexicana Woot. und
         A. Wrightii
         Reference(s):   1. Viehoefer, Capen, J.Amer.Chem.Soc. 45 〈1923〉, 1942, CODEN:
            JACSAT
INP   in den Blueten von Artemisia coerulescens L.
         Reference(s):   1. Herndlhofer, Mikrochemie 5, 21, CODEN: MIKRAJ Centralblatt:
            1927 I, 1992
INP   in den Blaettern von Artemisia maritima L. und A. monogyna
         Reference(s):   1. Pat. No.: 444850, D.R.P., Soteria Chem.-pharm. Fabr.
            Friedlaender: 15, 1524
            2. Pat. No.: 346947, D.R.P., Soteria Chem.-pharm. Fabr.
            Friedlaender: 13, 1065

L3   ANSWER 2 OF 22

CN    2-〈3,4-dimethoxy-phenyl〉-5-hydroxy-3,6,7-trimethoxy-chromen-4-one
      2-〈3,4-Dimethoxy-phenyl〉-5-hydroxy-3,6,7-trimethoxy-chromen-4-on

---> Handbook Data <---
Isolation from Natural Product:
INP   Isolierung aus Artemisia absinthium
         Reference(s):   1. Tunmann, Isaac, Arch.Pharm.(Weinheim Ger.) 290 〈1957〉 37, 42,
            CODEN: ARPMAS
            2. Cekan, Herout, Collect.Czech.Chem.Commun. 21 〈1956〉 79, 81, CODEN:
            CCCCAK
            3. Adrian, Trillat, Bull.Soc.Chim.Fr. 21 〈1899〉 234, CODEN: BSCFAS
INP   Isolierung aus Artemisia arborescens
         Reference(s):   1. Mazur, Meisels, Bl. Res. Coun. Israel 5 A 〈1955〉 67
```

Additional Search Hints

- The terms in many of these fields are in German:

Fields Containing German Terms	
BF	Biological Function
CPD	Crystal Property Description
ECOL	Ecological Data
INP	Isolation from Natural Product
PUR	Purification
TOX	Toxicity
RSTR	Related Structure
	Many note fields

- A stopword list is applied to all text search fields:

ABOUT	BOTH	IF	SEE	THROUGH
AFTER	BUT	*IN	SEEN	THUS
ALL	BY	INTO	SHOULD	TO
ALREADY	COULD	IS	SINCE	WAS
ALSO	DO	IT	SUCH	WE
ALTHOUGH	DOES	ITS	THAN	WERE
ALWAYS	DURING	MAY	THAT	WHAT
AMONG	EACH	MORE	THE	WHEN
AN	EITHER	MOREOVER	THEIR	WHERE
AND	FOR	MOST	THEM	WHETHER
ANY	FROM	MUST	THEN	WHICH
ARE	FURTHER	*NO	THERE	WHILE
*AS	HAD	OF	THEREFORE	WHOSE
*AT	HAS	ON	THESE	WILL
*BE	HAVE	ONLY	THEY	WOULD
BECAUSE	HAVING	OR	THIS	
BEEN	HERE	OTHER	THOSE	
BETWEEN	HOWEVER	OUR	THOUGH	

*The starred stopwords are also chemical symbols, and one is a chemical formula. These are: As (arsenic), At (astatine), Be (beryllium), In (indium), No (nobelium), and NO (nitric oxide). These stopwords have been indexed when they are used in the record as a chemical symbol or formula.

- Author names may be searched in the /AU field.

    - Only surnames are posted.
    - For less than 10 names, all names are searchable.
    - For 10 or more names, only the first name is searchable.

    ```
    => S KNOEVENAGEL/AU
    L3     161 KNOEVENAGEL/AU

    => D HIT

    L3   ANSWER 1 OF 161

    ---> Handbook Data <---
    Preparation:
    PRE
        Educt:   isatin, 4-amino-phenol
        Reag:    iodine, alcohol
        Reference(s):
        1. Knoevenagel, J.Prakt.Chem. ⟨2⟩89 ⟨1914⟩, 48, CODEN: JPCEAO
    ```

- Full abbreviated journal names are searchable in the /JT field.

    - Use EXPAND to find variations.

    ```
    => e E TERTA/JT 15
    E1              1 TERAHEDRON LETTERS/JT
    E2              1 TERRAHEDRON LETTERS/JT
    E3              0 TERTA/JT
    E4              1 TETARAHEDRON/JT
    E5              1 TETILBER./JT
    E6              1 TETRAHDERON LETTERS/JT
    E7              1 TETRAHDRON LETTERS/JT
    E8              1 TETRAHEDROM/JT
    E9           2743 TETRAHEDRON/JT
    E10             1 TETRAHEDRON 7/JT
    E11             1 TETRAHEDRON LETERS/JT
    E12          1014 TETRAHEDRON LETT./JT
    E13             1 TETRAHEDRON LETTER/JT
    E14             1 TETRAHEDRON LETTES/JT
    E15             1 TETRAHEDRON LETTTERS/JT

    =>
    ```

OTHER SEARCH INFORMATION
System Limits

- Truncation limit for text terms—2,000 unique stems

    Message: TRUNCATION LIMITS EXCEEDED - SEARCH ENDED

- Limit for numeric ranges—20,000 unique values

    Message: TRUNCATION LIMITS EXCEEDED - SEARCH ENDED

- Limits for structure searching:

SAMPLE (SSS, EXA, or FAM)	1,000 iterations 50 answers
FULL or RANGE (SSS, EXA, or FAM) (online)	20,000 iterations 5,000 answers
FULL or RANGE (SSS, EXA, or FAM) (batch)	50,000 iterations 10,000 answers

OTHER SEARCH INFORMATION V-6
Help

- Online HELP messages cover all aspects of search and display.
- STN Database Summary Sheet and STN Beilstein File Database Description provide information on search and display fields.
- For detailed help on database content, search strategy, and chemical issues, contact the Beilstein Institute in Frankfurt

> Beilstein Institute   Tel.: (+49 69) 791 7311
> Varrentrappstr. 40-42   (+49 69) 791 7492
> D-6000 Frankfurt/Main 90
> F.R. Germany

- For discounts on the database price, contact the database supplier.

> European and other countries:
> Ms. Griepke   Tel.: (+49) 6221-48757 (direct)
> Springer-Verlag GmbH & Co. KG   (+49) 6221-4870 (general)
> Dept. New Media
> Tiergartenstr. 17
> D-6900 Heidelberg
> F.R. Germany
>
> United States, Canada:
> Dr. R. Badger   Tel.: (212) 4601622 (direct)
> Springer-Verlag New York, Inc.   (212) 4601500 (general)
> 175, Fifth Avenue
> New York, N.Y. 10010, U.S.A.
>
> Japan:
> Mr. T. Hirano   Tel.: (03) 8120331 (general)
> Springer-Verlag Tokyo, Inc.
> 37-3, Hongo 3-chome, Bunkyo-ku
> Tokyo 113, Japan

- For general help, contact the nearest STN Service Center or the STN Representative for your country. Enter "HELP STN" online for a list of service centers.

VI. Practice Questions

1. IR data for a substance with a molecular formula of $C_6H_{11}NO$ suggest the presence of an amide group, probably a lactam. NMR data suggest three —CH2— groups and the ring fragment —C—C(O)—NH—C—. Have spectral data been reported for substances fitting these criteria?
2. A substance isolated from a species of *Lupinus* has a molecular formula of $C_{15}H_{22}N_{2O}$. The molecule contains two 6-membered rings, each containing 1 nitrogen atom, 5 carbon atoms, and a bridging methylene group. Have substances fitting this description been reported in the literature?
3. Find ways of preparing benzimidazoles from *o*-phenylenediamines.
4. Find ways of preparing gluteraldehyde.
5. Find oximes containing one halogen atom and with melting points between 200 and 225°C.
6. Find the structures for substances isolated from various species of *Digitalis*.
7. Find glucose derivatives with melting points of 175 ± 5°C for which ORD data have been reported.
8. Have any of the positional isomers of trimethyl quinoline (MF = $C_{12}H_{13}N$) been reported to have melting points greater than 75°C? (Hint: All positional isomers should have the same Lawson Number.)

VII. Appendices

Appendix A — Periodic Group Codes
Appendix B — Atomic Weights
Appendix C — Controlled Terms

Appendix A — Periodic Group Codes

Periodic group codes are generated for each element in the molecular formula, except carbon and hydrogen.

A1	A2		B3	B4	B5	B6	B7	B8			B1
Li	Be										
Na	Mg										
K	Ca	T1	Sc	Ti	V	Cr	Mn	Fe	Co	Ni	Cu
Rb	Sr	T2	Y	Zr	Nb	Mo	Tc	Ru	Rh	Pd	Ag
Cs	Ba	T3	La**	Hf	Ta	W	Re	Os	Ir	Pt	Au
Fr	Ra		Ac***								

B2	A3	A4	A5	A6	A7	A8
						He
	B		N	O	F	Ne
	Al	Si	P	S	Cl	Ar
Zn	Ga	Ge	As	Se	Br	Kr
Cd	In	Sn	Sb	Te	I	Xe
Hg	Tl	Pb	Bi	Po	At	Rn

**LNTH | Ce | Pr | Nd | Pm | Sm | Eu | Gd | Tb | Dy | Ho | Er | Tm | Yb | Lu |

***ACTN | Th | Pa | U | Np | Pu | Am | Cm | Bk | Cf | Es | Fm | Md | No | Lr |

SHEL | 104 Ung | 105 Unp | 106 Unh | 107 Uns | 108 Uno | 109 Une | 110 Uun | 111 Uuu |

APPENDICES
Appendix B — Atomic Weights

Symbol	Relative atomic mass	Symbol	Relative atomic mass	Symbol	Relative atomic mass
Ac	227.0278	Gd	157.2700	Po	210.0000
Ag	107.8680	Ge	72.5900	Pr	140.9077
Al	26.9815	H	1.0079	Pt	195.0800
Am	243.0000	He	4.0026	Pu	242.0000
Ar	39.9480	Hf	178.4900	Ra	226.0254
As	74.9216	Hg	200.5900	Rb	85.4678
At	210.0000	Ho	164.9304	Re	186.2070
Au	196.9665	I	126.9045	Rh	102.9055
B	10.8100	In	114.8200	Rn	222.0000
Ba	137.3300	Ir	192.2200	Ru	101.0700
Be	9.0122	K	39.0983	S	32.0600
Bi	208.9804	Kr	83.8000	Sb	121.7500
Bk	247.0000	La	138.9055	Sc	44.9559
Br	79.9040	Li	6.9410	Se	78.9600
C	12.0110	Lu	174.9670	Si	28.0855
Ca	40.0800	Lw	257.0000	Sm	150.3600
Cd	112.4100	Md	256.0000	Sn	118.6900
Ce	140.1200	Mg	24.3050	Sr	87.6200
Cf	251.0000	Mn	54.9380	T	3.0170
Cl	35.4530	Mo	95.9400	Ta	180.9479
Cm	247.0000	N	14.0067	Tb	158.9254
Co	58.9332	Na	22.9898	Tc	99.0000
Cr	51.9960	Nb	92.9064	Te	127.6000
Cs	132.9054	Nd	144.2400	Th	232.0381
Cu	63.5460	Ne	20.1830	Ti	47.8800
D	2.0141	Ni	58.7100	Tl	204.3830
Dy	162.5000	No	253.0000	Tm	168.9342
Er	167.2600	Np	237.0482	U	238.0289
Es	254.0000	O	15.9994	V	50.9415
Eu	151.9600	Os	190.2000	W	183.8500
F	18.9984	P	30.9738	Xe	131.2900
Fe	55.8470	Pa	231.0359	Y	88.9059
Fm	253.0000	Pb	207.2000	Yb	173.0400
Fr	223.0000	Pd	106.4200	Zn	65.3800
Ga	69.7200	Pm	145.0000	Zr	91.2200
Gd	157.2700	Po	210.0000		

Appendix C — Controlled Terms

Term	Code
Structural data Elementary analysis	CTGEN

Conformation
  Conformation
  Conformer equilibrium
  Energy difference between the conformers
	CTCFM

Skeletal characteristics
  Electron distribution
  Interatomic distances and angles
	CTSKC

Electrical polarizability
  Atom polarization
  Electron polarization
	CTELP

Coupling phenomena
  Electron–nucleus coupling constants
  ESR–hyperfine coupling constant
  Spin–spin coupling constants
	CTCPL

Molecular energy
  Bond energy
  Electron affinity
  Force constants
  Fundamental vibrations
  Proton affinity
  Resonance energy
	CTMEN

Crystal phase
  Association in the solid state
  Crystal growth
  Crystal habit
  Crystal morphology
  Crystal structure determination
  Interplanar spacing
  Polymorphism
  Rate of crystallization
  Rate of transition
	CTCRY

Liquid phase
  Association in the liquid state
  Radial distribution function
  Rate of evaporation
  Relaxation time for reorientation
  Rotational correlation time
  Self-association in solution
  Structure of the liquid
  Supercoolability
	CTLIQ

Appendix C — Controlled Terms

Term	Code
Gas phase	CTGAS

    Association in the gas phase

Mechanical properties	CTMEC

    Acoustic relaxation
    Adiabatic compressibility
    Elasticity constants
    Internal pressure
    Isothermal compressibility
    PVT relationship
    Sound absorption
    Specific volume
    Velocity of sound
    Virial coefficients of the equation of state
    Volume change on melting

Calorific data	CTCAL

    Cryoscopic constant
    Ebullioscopic constant
    Enthalpy
    Enthalpy of self-association
    Entropy of melting
    Entropy of other phase transitions
    Entropy of sublimation
    Entropy of vaporization
    Heat of combustion at constant volume

Optical data	CTOPT

    Cotton effect (abnormal curve)
    Crystal refractive indices
    Degree of depolarization of Rayleigh scattering
    Diffraction
    Electrical birefringence (Kerr effect)
    Iso- and anisotropic components of Rayleigh scattering
    Linear dichroism
    Magnetic birefringence (Cotton–Mouton Effect)
    Magnetic circular dichroism
    Magnetorotation
    Mechanical birefringence
    Mutarotation coefficient
    Natural birefringence
    Photochromism
    Plain curve
    Rayleigh scattering
    Reflection
    Thermochromism

Appendix C — Controlled Terms

Term	Code
NMR data	CTNMR
CIDNP	
Dynamic NMR	
INDOR	
Linewidth of NMR absorption	
NMR in liquid crystal phase	
NMR with shift reagents	
NOE	
Second moment of NMR absorption	
Spin–lattice relaxation time T_1	
Spin–spin relaxation time T_2	
Two dimensional NMR	
NQR data	CTNQR
Nuclear quadrupole resonance	
ESR data	CTESR
CIDEP	
ELDOR	
ENDOR	
ESR absorption	
ESR spectrum	
g-Factor	
Rotational spectrum	CTROT
Intensity of microwave bands	
Microwave spectrum	
Rotational spectrum	
Rotational–Raman spectrum	
Stark effect	
Vibrational spectrum	CTVIB
Anisotropy of IR bands	
Degree of depolarization of Raman bands	
Fermi resonance	
Fine structure of IR bands	
Intensity of IR bands	
Intensity of Raman bands	
Polarization of IR bands	
Reflection spectrum in the IR region	
Resonance–Raman effect	
Electronic spectrum	CTESP
Anisotropy of absorption bands	
Oscillator strength	
Reflection spectrum in UV/VIS region	
Singlet–triplet absorption	
Solvatochromism	
Triplet–triplet absorption	
Vacuum UV spectrum	

Appendix C — Controlled Terms

Term	Code
Emission spectrum	CTEMS

    Degree of polarization of the emission
    Electroluminescence spectrum
    Emission quenching
    Emission spectrum in the infrared region
    Lasing properties
    Lifetime of the excited state
    Quantum yield of emission
    Radioluminescence spectrum
    Sonoluminescence spectrum
    Thermoluminescence spectrum
    Triboluminescence

Other spectroscopic methods — CTOSM
    Appearance potential
    ESCA
    Ion cyclotron resonance
    Mössbauer effect
    Negative ion spectroscopy
    Photoelectron spectrum

Magnetic data — CTMAG
    Anisotropy of magnetic susceptibility
    Magnetic moment

Electrical data — CTELE
    Angle of dielectric loss
    Anisotropy of dielectric constant
    Cole–Cole diagram
    Critical frequency (or wavelength)
    Dielectric increment
    Dielectric loss
    Dielectric relaxation time
    Dielectric saturation
    Dielectric strength
    Electrical conductivity
    Photoconductivity
    Photoelectricity (Becquerel effect)
    Piezoelectricity
    Relaxation frequency
    Thermoelectricity

Appendix C — Controlled Terms

Term	Code
Electrochemical behavior	CTECB

    Acidity
    Acidity function H_0/H_r
    Autoprotolysis
    Basicity
    Collision cross-section
    Electrolytic dissociation
    Enthalpy of neutralization
    Ionization cross-section
    Kinetics of dissociation
    Oxidation potential
    pH of aqueous solutions
    Polarographic current–voltage curve
    Protonation equilibrium
    Stability constant
    Thermodynamic parameters for autoprotolysis
    Thermodynamic parameters of dissociation
    Volume change on dissociation

Term	Code
Mass spectrum	CTMS

    Fragmentation
    Molecular peak

Term	Code
Solution behavior (MCS)	CTSOLM

    Dissolving capacity
    Miscibility
    Mutual solubility
    Rate of dissolution
    Solubilizing

Term	Code
Liquid–liquid systems (MCS)	CTLLSM

    Critical solution temperature
    Distribution between solvents 1 and 2
    Equilibrium of liquid phases
    Liquid–liquid phase diagram
    Solution equilibrium
    Temperature of separation

Term	Code
Liquid–solid systems (MCS)	CTLSSM

    Eutectic
    Liquid–solid phase diagram
    Melting diagram
    Solidification diagram
    Solidification points of mixtures

Appendix C — Controlled Terms

Term	Code
Liquid–vapor systems (MCS)	CTLVSM
Activity coefficients of compound in mixture	
Boiling point diagram	
Boiling points of mixtures	
Critical data for mixtures	
Liquid–vapor equilibrium	
Liquid–vapor phase diagram	
Partial pressures of the components	
Vapor pressure diagram for the mixture	
Gas phase system data (MCS)	CTGASM
Adiabatic compressibility	
Isothermal compressibility	
Partial molar volume	
PVT relationship	
Virial coefficients	
Volume change on mixing	
Transport phenomena (MCS)	CTTRAM
Diffusion	
Thermal diffusion	
Viscosity	
Energetic data (MCS)	CTENEM
Enthalpy of dilution	
Enthalpy of evaporation	
Enthalpy of mixing	
Enthalpy of mixtures	
Enthalpy of solution	
Entropy of mixtures	
Excess thermochemical parameter	
Heat capacity of mixtures	
Boundary surface phenomena (MCS)	CTBSPM
Contact angle with compound	
Interfacial tension	
Pressure–surface isotherm	
Spreading pressure	
Surface moment	
Surface potential	
Surface tension	
Adsorption data (MCS)	CTADSM
Adsorption	
Adsorption isotherm	
Chemisorption	
Desorption	
Enthalpy of adsorption	
Further physical properties of the adsorbed material	

Appendix C — Controlled Terms

Term	Code
Association data (MCS)	CTASSM

    Association with compound
    Dipole moment of the complex
    Enthalpy of association
    Further physical properties of the complex
    Spectrum (e.g. UV, IR, NMR) of the complex
    Stability constant of the complex with ...

Term	Code
Unchecked data	CTUNCH

    Boundary surface phenomena
    BPC
    Compressibility
    Conformation
    Electrical properties
    Electrochemical properties
    ESR
    Fluorescence
    Heat capacity
    IR
    Liquid crystal properties
    Luminescence
    Magnetic properties
    NMR
    Optical properties
    Phosphorescence
    Polarography
    Raman effect
    Rotational correlation time
    Solid state structure properties
    Thermodynamic properties
    Thermodynamic properties of system with ...
    Ultrasonic properties
    UV/VIS
    Viscosity

If you have any concerns about our products,
you can contact us on
ProductSafety@springernature.com

In case Publisher is established outside the EU,
the EU authorized representative is:
**Springer Nature Customer Service Center GmbH
Europaplatz 3, 69115 Heidelberg, Germany**

Printed by Libri Plureos GmbH
in Hamburg, Germany